SCREEN SLAVE

Why and How We Should Free Ourselves From Our Devices

By

ANTHONY T.J. JOHNSON

This book is dedicated to my mom, Ruth Lynn Johnson; God rest her soul. It was my mom who made it clear to me that I was a screen slave and she helped me realize the importance of engaging with her while in her presence. Every second with each other matters because when your time is up, there's nothing you can do about it.

TABLE OF CONTENTS

INTRODUCTION

Something disturbing is happening all across the globe, destroying minds and relationships and making the world a more dangerous place. I am speaking of something most of us use daily–our smartphones. Just look around! Everywhere you go, people are walking with their necks bent over, glued to their phone, being dumbed down into walking-dead zombies. And many of those who are not walking are texting and driving, causing many unnecessary accidents and deaths on the road–multi-tasking is great when preparing a meal, but not so great when driving.

Some will say they are not a slave to their phone; they are in full control. An addict is usually in denial at first

1

when confronted with their issue. For me, being a screen slave is when you have no control or freedom from your phone's hold over your life. Let me ask you a question: if your phone made an alert sound right now, would you stop immediately to look at it? Of course, you would. You cannot help it, right? It is like we have to put the world on pause to stop and check our phones. It consumes our minds, actions, and entire lives, causing sleepless nights and deteriorating psychological and health issues.

Hopefully, after you read this book, you will begin to take back control of your life by not allowing a handheld device to dictate your actions, thoughts, and emotions. I have seen it firsthand in my life, and with my wife, kids, family, and friends. Honestly, I am tired of it! It has damaged my relationships in the past because I put more time into my phone than the person I was with.

I recall going to visit my mom when she said, "Every time you come over you are on your phone. Put your phone away, you are here to visit me!" From that day forward, whenever I was with my mom, I was with my mom and not my phone. Isn't that foolish? Putting a phone over a person... I could not help it at the time, I was so consumed and blinded by a false reality of fake importance that I actually lost or almost lost some important people and things in my world that really matter. I was a Screen Slave.

Enough is enough!! Today is the day you begin to break free from your phone addiction and begin a new life of clear control and fresh energy.

Day 1

VALUE OF ENGAGEMENT

Having survived or overcome my screen addiction, I can now clearly see that most people do not engage with each other anymore–especially with loved ones. I wrote this book because I want to see people walking down the street together, talking and holding hands, not walking side by side holding their phones. I want to see families together again at the dinner table, laughing and discussing life, not consumed by things that are unimportant.

The addiction might be so small you don't even notice it. For instance, I was addicted to my emails, but I wasn't aware until my wife brought my attention to it. I was in denial at first, but had to step back and reflect on it. I was addicted to checking and deleting my emails. Plus,

I have maybe ten email accounts, so you know it was time-consuming. I would say 90% of what I was distracted by were unimportant emails and or junk mail. It's always the unimportant and inconsequential things that take your time away from the things that really should matter in our lives.

End of Day 1: Think Free Thoughts

Take a walk today. Try to get 30-60 minutes in. Or you can do 30 minutes today and 30 minutes another day this week. Turn the phone off and take a notebook with you to write the things you see and the sounds you hear.

Notes

Day 2

REAL-TIME
MEMORY MOMENTS

It is time to start creating real-time moments in real-time. I recently went on vacation with my wife to celebrate our 4-year anniversary. We were finally alone for a few days with no kids... I said no kids, lol, can I get an Amen? I love my kids! I enjoy being their father and spending time with them daily, but one day they will leave home, and wifey and I will have to look into each other's eyes as friends and lovers or strangers with no benefits. That is another book for another time, but you get where I am going with this; I must continue to work on my marriage if I want it to work. The important thing to consider here is that this was the first time we did not post our every move or detail about the trip on social media. We waited to share our love experiences with the world until we got home from our love vacation.

Even my wife said she could tell the difference between creating real-time moments and memories. You actually get to experience the authenticity of it all and not just look back at pictures and videos and not really remember doing this or that. Now how does that make you feel? Not good at all, I'm sure, but today is the day you begin to make that paradigm shift. Every moment is not a Kodak moment or a featured film clip. The *real* moment is engraved in your mind in real-time.

End of Day 2: Think Free Thoughts

Stay in the moment this week–no posting on social media! I know this may seem impossible, but trust me... when you do make it back to social media, it will be like you never left. You could leave for a whole calendar year and things would be the same or worse. No one really misses you because they are not your true family or friends, it is all a facade. You can take pics for memories–but no posting! Do your best to stay in the moment with real-time memories that cannot be duplicated. There's nothing worse than looking back at a picture or video when you cannot even remember that moment.

NOTES

$\mathcal{D}$ay 3

START EACH DAY WITH THE RESET BUTTON

It is a brand-new, beautiful day that God has given us. Some wake up using an alarm, like me, and others have an internal clock that tells them to get up. Either way, most of us grab our phones before saying good morning to our significant other or even taking a moment to thank the Most High for the new day and new opportunities ahead. Instead of grabbing our phones first thing in the morning to get caught up in the web of lies from social media, celebrity drama, and gossip; instead hold down the power button on your phone until the restart button appears and press it. This does a couple of things; it allows our phones to clear out some things that are clogging its memory and ability to function at a high level. As well, the reset ensures the first moments of your day are not filled with unnecessary negative energy and allows you some moments to tap into your higher self as you restart the awareness of the higher source of life in your day and mind. This gives you an advantage that many do not have because they do the opposite and are led throughout the day by the negative energy they feed their mind and body. If you have positive energies fed and flowing at the start of the day, it will give you spiritual clarity, divine insight, and creative capabilities on a higher frequency because the start of the day is vital to successful minds.

End of Day 3: Think Free Thoughts

Go to a coffee shop or restaurant and leave your phone in the car. Observe how other people are interacting with each other. Is there eye contact? Is everyone on their phones, not fully engaging with each other? Do you see yourself in others? How does it make you feel?

NOTES

Day 4

How Did We Start
Following the Crowd?

This one is mind-blowing, even for me, because I found myself following the crowd, and it was not my intention to do so at all. I do not know where it started for me. All I know is that, one day, I woke up and had an epiphany... I asked myself why I had been sharing my every move on social media. Why did it matter if I told everyone I'd just entered the coffee shop? Or post photos of what I was eating and drinking that day? If you are not aware of what you are doing consciously, then you open yourself to being manipulated subconsciously, blowing in the wind with every trend and challenge. I recently heard that the makers of TikTok do not allow their children to use the app or any other social media apps. As a matter of fact, TikTok in China is vastly different than here in America. Their users get videos on the arts, math, science, and things of that nature. While we get silly dances, crazy videos, and no-benefit challenges that dumb us down, not adding value or strengthening our minds.

End of Day 4: Think Free Thoughts

Uninstall all social media apps on your phone for a minimum of 3 days. I say uninstall, because if you do not uninstall, you will not be able to resist going on it, because, at this moment, the addiction is too strong; but things are changing for the better, starting today.

NOTES

Day 5

THE PRICE OF
COMPARISON

One of the hardest things to overcome is the trap of social media comparison. Before the internet, "keeping up with the Joneses" meant just competing with your neighbors for the biggest house, newer cars, clothing, etc. Now, with the world wide web, if we are not making the conscious decision to be aware of the traps, we will find ourselves competing with the entire world, which is very detrimental to our health spiritually, physically, mentally, and financially. Not to mention it is impossible! Listen, what you see is not real life; people only show you their so-called highlights. Everyone shows themselves at the party, living it up, but never at the end of the party, throwing it up.

End of Day 5: Think Free Thoughts

Social media has a way to make us feel ungrateful for where we live, the car and family we have, etc. Today, set an alarm every two hours and, each time, write down two things you are grateful for and how you can express that gratitude.

NOTES

Day 6

COMMUNICATION
LEADS TO APPRECIATION

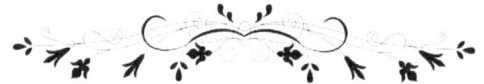

I want you to understand that when you have a conversation with someone, if you take a moment inside the conversation, you can have a moment of appreciation. For example, I recently attended a wine tasting with my wife. In the midst of the environment, I looked around and noticed the wide range of ages enjoying the wine flavors of life. Those in their twenties were conversing with those in their seventies. Two different generations had the commonality of enjoying a glass of wine or two or more, relating on the same level, discussing palate flavors, sophisticated benefits, and lifestyle events. No phones in hands, just a glass of wine along with smiles and laughs. If a phone was out, it was someone showing pictures of their family.

End of Day 6: Think Free Thoughts

Today, ponder on a new group activity or networking event to attend. It's time to meet new people and expand your thinking. To achieve more, you have to explore more, especially outside your comfort zone.

NOTES

Day 7

PROCRASTINATION AND EXAGGERATION

One of the biggest hindrances to your progression is procrastination! Trust me, I definitely understand, because I struggle with it at times myself. Something important that needs to be done in a timely manner gets pushed to the side in our minds. We think we will get to it soon, but it never happens, and if it does happen, it is rushed, and it is not the best you have to offer. Our phones subconsciously create a false narrative that we can spend just a little time on it and still have time to complete our task at hand.

You have to tune in and take control of your thoughts and focus. There is nothing worse than ending your day and realizing very little or nothing you had planned got accomplished. After a while, days turn into weeks, weeks to months, and months into years. As far as procrastination goes, going online will only exaggerate everything, with streaming service movies and shows, addictive games, or social media apps, it's easy to procrastinate. Most of it is not real, even though they go under the banner of 'reality TV.' All of these distractions will keep you unfocused. Have you been caught up in a mess longer than you want to? Do you fill your mental space with negativity that clouds your good judgment abilities?

End of Day 7: Think Free Thoughts

Today is another day for taking a bold step to regain control of your life and not allow a device to control and dictate your thoughts and actions. For 1 week, turn off the TV and all social media apps. It is a big step, but even bigger freedom choices are ahead. Keep moving forward... 1 step at a time... 1 day at a time.

NOTES

Day 8

DAMAGE CONTROL

Addiction to a phone is certainly something you have seen or heard about before, it is no secret that we have become dependent on our devices. Smartphones are slowly replacing face-to-face interactions in our relationships, and every aspect of our lives appear to revolve around the usage of our smartphones. As the number of individuals using mobile phones rises, so does the number of people who become dependent on them.

The original purpose of cell phones was to make people's lives simpler.

These devices were created to help us, not to take over our functions, they were not made to be a substitute for human interaction. Without limits, even life-saving technology like a cell phone can be misused. This is gradually resulting in self-induced isolation, which gives rise to mental illnesses such as depression in return. While cell phones are wonderful and highly convenient to use, their excessive usage can have a negative impact on society and people's quality of life. As a result, a lot of parents are looking for apps and other things that they can use to help their children deal with the challenges presented by modern life innovations. This is a major concern.

According to a Nielsen study, more than 70% of kids under the age of twelve own tablets or iPads. According to research published in the *Journal of Pediatrics*:

1. 20% of one-year-olds own a tablet.
2. 28% of the surveyed 2-year-olds could use a mobile device independently.
3. 28% of parents use a mobile device to put their children to sleep.
4. Tablet and smartphone use in young children has devolved into child's play, with children able to touch and swipe efficiently before they can walk or talk.

5. More than half of US citizens cannot go a day without their mobile devices, with many of us checking our phones first thing in the morning before getting out of bed; our children are no different. Children mimic their parents' actions.

Therefore, the decisions we make right now will influence the younger ones.

My mother, whom I hold in very high regard, brought up my brother and me as a single parent. My determination to make adjustments to my screen time habits was strengthened when she told me that I should not visit if I could not commit to having conversations with her during those visits. I remember her correcting me on how much time I spent on the phone. That was a wake-up call for me, and as a result, I made it a habit to ensure that whenever I went to see my mother, I gave her my utmost attention and made sure that I was fully engaged in the conversations that we had.

Before this development, I would use my phone for the entirety of each day, spending at least ten hours on it, and I would do this even while I was working because I was so engrossed in what I call the Matrix of Social Media; just lost in the sauce I was being fed. My interpersonal relationships with others had been affected as well. Sometime around 2008, I had recently joined Facebook and I found out that I was able to connect to new people online on a daily basis; it was an incredible experience. This was both a blessing and a curse because it resulted in the decline of my most valued relationships back then. While this was not the only reason the relationship ended, it was a contributing factor. I was not paying attention to what was being said during our interactions. In many of our exchanges, she insisted that I maintain my concentration. This was a challenge for me. When I think back on it now, I realize that was one of the things that contributed to the void that was created in the relationship. I could have done better by, for example, being engaged in conversations and being the companion she needed at the time.

Moving on, it was not until I was on the receiving end of one of these one-sided conversations that I realized how damaging and hurtful they are. When someone else pointed out the obvious to me, reality smacked me right in the face and it could not be ignored any longer. I took my mother's advice and began making changes in my behavior, such as deciding not to bring my phone with me when I was with other people and making an effort to participate actively in conversations rather than checking my phone.

In summary, electronic technologies such as mobile phones and other means of communication have displaced the intimacy that comes from face-to-face interactions with others. Human relationships and personal interactions are steadily decreasing. Think back to the activities you used to appreciate before you were glued to your phone all the time, like watching a show, spending time with your family, or taking in the countryside on a trip. What did your family enjoy doing? With determination, we can all gain back the affections and intimacy that we have lost as the result of incessant usage of technological devices.

End of Day 8: Think Free Thought

Try this today... When there are two or more people, ignore your phone. Whether you are with a group of family or friends, or with just one other person, put your phone on silent to help resist the temptation to check the device while in the presence of another. Do this until it becomes second nature, when you give those around you, all of you.

NOTES

$\mathcal{D}$ay 9

ATTRACTION TO

DISTRACTION

For me, the moment I realized that smartphones are slowly robbing us of meaningful interactions with other people came when I understood that my relationships with my family and friends should take precedence over any notifications. These notifications are intended to keep you interested and to draw you deeper into the screen slave matrix. There will be advertisements in addition to other things to keep our interest. Not only are we more likely to use our phones for longer periods, but the notifications and advertisements that appear on our screens can also lead us to make purchases of things that we do not necessarily require at the moment or possibly any moment in life.

At the time, I was unable to resist giving into the temptation of using my phone to pass the time, which made me realize that I was becoming a part of a matrix. It was like me and my wife were slowly being consumed with the urgency to use our phones at times where face-to-face communication was needed, which is when I knew that I had to wake up and something had to change.

In addition, texting is the function of smartphones that many use more frequently, but it reduces the number of meaningful interactions people have with one another and also increases the likelihood that they will misunderstand one another.

Tone can often be misinterpreted via text. There are some scenarios in which you might be able to send a text message, but you are fully aware that it has an extremely low probability of being well received. There is always the possibility that you will be misunderstood. You may need to add more substance because the tone of the message could be misunderstood, as opposed to a phone call. Sometimes, if you do not add a smiley face or a laughing emoji, it can turn into an ugly outcome.

When you allow yourself to become a slave to your screen, you allow yourself to become distracted from the things that actually matter in your life. We are all guilty of this. What matters now is changing the narrative.

In addition, because of the widespread adoption of modern technology, everyone, from young toddlers to elderly people, has instant access to an apparently limitless amount of information and tools. We were not designed to spend our days engrossed in front of a screen, whether it be a computer monitor, TV, tablet or a mobile phone. As smartphones evolve to perform more complex tasks and provide more assistance, their users will inevitably come to rely more heavily on these devices in their day-to-day lives. However, make sure you do not lose sight of your family and spend some quality time with them.

Unhealthy trends on social media, such as the trend of teenagers wearing shirts that say "Antisocial," are one of the many ways that social media poses a threat to the healthy way of living. This trend is an example of how social media can threaten healthy lifestyles. We were not created to be anti-social, but the exact opposite–to be social and have human connection and love for one another.

It is impossible to build a strong community or lead a happy life without engaging in conversation and social activities with other humans. People need people, and engaging in the community helps build social skills. No one should be alone. The enemy comes to isolate us and divide us, which leads to anger, depression, anxiety, and other detrimental psychological and social issues. Think about animals in nature, the predator never goes after prey connected with the group and community, but always goes after prey that is disconnected and not aware. We all know what happens to the animal that is lost and isolated; I do not need to explain any further. It never ends well with isolation and disconnections.

End of Day 9: Think Free Thought

Right now is a good time to decide if you want to take the "red pill" or "blue pill" which is a reference from a scene out of the 1999 hit film the Matrix. A choice between the willingness to learn a potentially unsettling or life-changing truth by taking the red pill or remaining in contented ignorance with the blue pill. Life is all about choices and what better way to learn about how social media was created and designed to enslave us mentally from those who help create it. Watch "The Social Dilemma" , a documentary on Netflix. Then watch "Childhood 2.0" – Social Media Documentary from Bark on YouTube. Around 3 hours of your time to see how addictive it was created and the dangers many families are dealing with mentally and make the red or blue pill decision for yourself.

NOTES

Day 10

REMOVE DISTRACTIONS

Changing this narrative is not going to be an easy thing to do as an adult, and it is going to be even more difficult to do as a teenager, which is why we have to set things straight by showing our kids some tough love. My eldest child is currently seventeen years old, which means she is in the prime of her teenage years, however, my wife and I have reinstated parental control settings on her phone after removing it some time ago.

The effects of removing the parental controls on the phone were clear to see because it affected her grades, two semesters in a row, so we had to do something about it. Now there is a limit of seven hours of screen time daily, and after the lapse of that time, she cannot text or use the internet or apps. She can only make phone calls.

These parental control settings are primarily focused on websites, screen time, and potentially harmful activities that could be used to lure children. This is the essence of accountability; you must accept responsibility for the consequences of your actions. If you are doing something incorrectly, you need to devise a strategy to fix it. Something needed to be adjusted, and that is what my wife and I did. She is a good student who is trying her best, but those adjustments are what made the difference, despite her having major issues with it–and with us.

Who can you trust to keep you accountable? Maybe you cannot trust yourself enough. You can also use your phone applications to help you set limits on your usage of your favorite social media apps. Stick to it. Unless you are getting paid. I suggest two to four hours daily is more than enough. Even if you are getting paid, work, then cut it off. Try working for two hours and taking a break for 30 to 60 minutes, and then working for another two hours and repeat. Spend that time with family and friends, moments go so fast, and you cannot get those moments back again.

It is hard to keep yourself accountable. You will say I will only be on here for one hour; but then you begin to waste your minutes and now it's turned into hours. So the best thing to do is to put your devices down. Samsung and iPhones both have options where you can put limits on individual apps. Let the devices help you set the limits on how much time you are spending on them. And then, take a break in between what you are doing, spend that time with your family and friends and engage with the locals in your community as well. How many neighbors do you know by name or even have casual conversations with? Stop just going into your home and not waving or conversing with those around you.

I've definitely tried to check in with my wife more often. She works and makes money on her phone, but I told her she still has to pause from taking customers' orders and use that time with her family, which could be reading a book, playing games together, watching shows, or anything you think of that involves engagement and laughter. At this point, she definitely knows the benefits, and she is working on it. I'm proud of her progression.

The challenge is to work on yourself, it is not enough to set limits, when we set the limits, we can still overwrite, we can go in there, and turn it off. She has done that as well and I tease her, saying I will set up parental control on the phone so I can cut it off at any time I feel it's needed, lol.

End of Day 10: Think Free Thought

Delete social media apps and all the apps you don't use! The best way to resist temptation is to remove the temptation. Take some of your power back today. Go through your phone and delete any app you have not used within the last 30-60 days. By doing this, you'll feel the power and freedom of deciding what to have or not to have on your phone. Just because you downloaded it does not mean it has to stay on your phone forever. Plus, your phone will work better, and not be cluttered up with useless apps taking up storage space.

NOTES

$\mathcal{D}ay\ 11$

REST TO BE YOUR BEST

If you find yourself unable to fall asleep at a reasonable hour and are tempted to stay up to engage in online conversation until the early hours of the morning, you may have a problem. Let me correct myself, you have a big problem! Looking at your phone when you should be asleep causes the body to believe it still needs to stay up, even though it is ready to shut down. Why can't we just turn off the lights and put our phones away to get the proper rest our bodies and minds deserve? According to SCL Health, the blue light emitted by your cell phone screen restrains the production of melatonin, the hormone that controls your sleep-wake cycle (aka circadian rhythm). This makes it even more difficult to fall asleep and wake up the next day. The circadian rhythm seems to be especially sensitive to blue light since it has a short wavelength. Studies also show that exposure to blue light can cause damage to your retinas. Also, according to the US Center for Disease Control and Prevention, one-third of adult Americans have a sleep disorder that can seriously affect their health.

To be clear, we all have to do better and make conscious efforts to limit our phone usage before and during bedtime hours. Doing so will not only save your mental stability, but also improve your health, which of course is vital for longevity.

End of Day 11: Think Free Thought

Many of us cannot sleep without white noise. We must have the TV or music turned on, or if you are like my wife, a fan blowing every night. I do not like it, but that's married life–compromising, lol. Try sleeping in peace for a change for the next week. Something new can lead to relaxing your mind and body fully without possibly negative words and sounds being forced upon your spirit. Who knows, you may actually hear an airplane flying, cars rolling down the street, or birds chirping and singing!

NOTES

Day 12

THE POWER OF MINDFULNESS

In a world where our attention is constantly being pulled in different directions, mindfulness can be a powerful tool to regain control. Mindfulness is the practice of being fully present and engaged in the current moment, without distraction or judgment. It is about noticing where we are and what we are doing, without being overly reactive or overwhelmed by what is going on around us.

When we are constantly checking our phones, we are not being mindful. We are not fully present in our own lives. We are missing out on the richness of the world around us and not giving our minds the chance to rest and recharge.

According to a study published in the *Journal of Behavioral Addictions*, individuals who reported higher levels of mindful attention and awareness also reported less problematic smartphone use. This suggests that mindfulness could be a valuable tool in combating screen addiction.

End of Day 12: Think Free Thoughts

Today, try practicing mindfulness. Set aside a few minutes to sit quietly and focus on your breath. Notice the sensation of the air entering and leaving your body. If your mind starts to wander, gently bring it back to your breath. Try to do this at least once a day and see if you notice any changes in your relationship with your phone.

NOTES

Day 13

THE ART OF MINDFULNESS

Mindfulness is the practice of being fully present in the moment, aware of where we are and what we are doing, without being overly reactive or overwhelmed by what's happening around us. It is a simple concept, but it's one that can have profound effects on our mental well-being.

In the context of our relationship with our phones, mindfulness can help us to use our devices more intentionally, rather than mindlessly scrolling through social media or checking our emails every few minutes. By being mindful, we can choose to engage with our phones in a way that serves us, rather than allowing our phones to dictate our behavior.

End of Day 13: Think Free Thoughts

Today, try to practice mindfulness when using your phone. Before you pick it up, ask yourself: Why am I picking up my phone? What do I intend to do with it? Is this the best use of my time right now? You might find that this simple act of mindfulness helps you to use your phone less and enjoy your life more.

NOTES

Day 14

THE POWER OF
UNPLUGGING

In our hyper-connected world, the idea of unplugging, disconnecting from our devices and the internet, can seem daunting. However, taking regular breaks from technology can have significant benefits for our mental and physical health.

Research has shown that excessive screen time can lead to a range of health issues, from eye strain and headaches to sleep disturbances and increased risk of mental health problems. By unplugging, even for a short period each day, we can give our minds and bodies a much-needed break and improve our overall well-being.

End of Day 14: Think Free Thoughts

Today, set aside a period of time, it could be an hour or even just 15 minutes, where you completely unplug from technology. Turn off your phone, close your laptop, and do something that does not involve screens. You could read a book, go for a walk, or simply sit and enjoy the silence. Notice how you feel during and after this period of unplugging. At first, you may have withdrawals, but over time, you will only notice the power, freedom and self control you have now discovered that has led to you choosing to think freely.

NOTES

Day 15

The Impact of
Notifications

Notifications are a constant source of distraction in our lives. Each time our phone buzzes with a new message, email, or social media alert, our attention is pulled away from whatever we were doing. This constant interruption can lead to increased stress and decreased productivity.

A study by the University of California, Irvine, found that it takes an average of 23 minutes and 15 seconds to get back to the task at hand after an interruption. Over the course of a day, these interruptions can add up to a significant amount of lost time and productivity.

End of Day 15: Think Free Thoughts

Today, try turning off non-essential notifications on your phone. This might include social media alerts, news updates, or any other notifications that are not critical to your work or personal life. Notice how it feels to have fewer interruptions throughout your day.

NOTES

Day 16

THE ILLUSION OF
MULTITASKING

Many of us pride ourselves on our ability to multitask, especially when it comes to juggling multiple digital tasks at once. However, research suggests that multitasking, especially with digital devices, is not as efficient as we might think.

A study by Stanford University found that heavy multitaskers–those who frequently switch between different types of media–performed worse on various cognitive tasks than those who preferred to complete one task at a time. The researchers concluded that multitasking can lead to decreased productivity and increased susceptibility to distractions.

End of Day 16: Think Free Thoughts

Today, try to focus on one task at a time rather than juggling multiple tasks at once. Notice how it feels to give your full attention to a single task and observe whether it impacts your productivity or stress levels.

NOTES

Day 17

THE ROLE OF SMARTPHONES IN ANXIETY AND DEPRESSION

There's growing evidence to suggest that excessive smartphone use can contribute to mental health issues like anxiety and depression. A study published in the journal *Computers in Human Behavior* found that people who use their phones more frequently are more likely to experience symptoms of anxiety and depression.

The constant connectivity that smartphones provide can lead to feelings of overwhelm and stress. Additionally, the pressure to maintain an active social media presence and the tendency to compare oneself to others online can contribute to feelings of inadequacy and low self-esteem.

End of Day 17: Think Free Thoughts

Today, take some time to reflect on how your smartphone use might be impacting your mental health. If you notice that you often feel anxious or stressed after spending time on your phone, consider setting boundaries around your phone use to protect your mental well-being.

NOTES

Day 18

THE IMPACT OF SMARTPHONES ON PHYSICAL HEALTH

While the mental health implications of excessive smartphone use are concerning, it is also important to consider the physical health effects. Prolonged smartphone use can lead to a variety of health issues, including eye strain, poor posture, and even physical pain.

According to a study published in the journal *Musculoskeletal Science and Practice*, frequent smartphone users are more likely to experience neck, shoulder, and upper back pain. This is often due to the poor posture that people adopt while using their phones, such as hunching over or craning their necks forward.

End of Day 18: Think Free Thoughts

Today, pay attention to your posture while using your phone. Try to hold your phone at eye level to avoid bending your neck and take regular breaks to stretch and move around. Consider using a stand or holder to support your phone if you are using it for extended periods.

NOTES

$\mathcal{D}$ay 19

THE POWER OF
DIGITAL DETOX

In our hyper-connected world, the idea of a digital detox–taking a break from all electronic devices–can seem daunting. However, research suggests that it can have significant benefits for our mental and physical health.

A study from the University of Pennsylvania found that limiting social media use to 30 minutes per day resulted in significant reductions in loneliness and depression. Similarly, a study from the University of East Anglia found that participants who took part in a digital detox reported feeling more productive, having better sleep, and experiencing improved relationships.

End of Day 19: Think Free Thoughts

Consider trying a digital detox for a day, or even just a few hours. Use this time to engage in activities that you enjoy and that are good for your well-being, such as reading a book, going for a walk, or spending time with loved ones. Pick up a new hobby or dust one off that you have put on the shelf.

NOTES

Day 20

THE ROLE OF MINDFULNESS IN BREAKING FREE FROM SCREEN ADDICTION

Mindfulness, the practice of being fully present and engaged in the current moment, can be a powerful tool for breaking free from screen addiction. By becoming more aware of our thoughts, feelings, and actions, we can make more conscious decisions about our smartphone use.

A study published in the journal *Mindfulness* found that participants who completed a mindfulness-based smartphone use reduction program showed significant decreases in daily smartphone use and improvements in self-reported mindfulness, well-being, and productivity.

End of Day 20: Think Free Thoughts

Oxygen or Wifi??

Too many of us are so addicted to our screens we can't even imagine our lives without it. Many parents have heard from a teenager they'd die if they don't have their phones. Really? Will we really die without our screens or are there more important things keeping us alive that we have forgotten or taken for granted? Today be mindful and meditate on things that really matter most and give life and meaning. No phone needed...just a paradigm shift with a new perspective for a new day.

NOTES

THE JOURNEY
TO FREEDOM

Congratulations! You have made it to Day 21! Over the past three weeks, you have taken significant steps towards breaking free from your screen addiction. Remember, this is a journey, not a destination. It is about progress, not perfection.

A study from the University College London suggests that it takes an average of 66 days to form a new habit. So, while you have made great strides in the past 21 days, it is important to continue practicing these new behaviors to solidify them into lasting habits.

End of Day 21: Think Free Thoughts

Reflect on the past 21 days. What changes have you noticed in your mood, productivity, relationships, and overall well-being? What strategies have been most helpful for you? Moving forward, how can you continue to prioritize real-life connections over screen time?

Remember, the power to change is in your hands. You have the ability to control your screen use, not the other way around. Here is to a future of clear control, fresh energy, and meaningful engagement with the world around you.

NOTES

Conclusion

As we reflect on this transformative journey, it is crucial to remember that breaking free from screen addiction is not a one-time event, but a continuous process. The past 21 days have been about laying the foundation for a healthier relationship with technology.

The exercises, reflections, and strategies we have explored are tools you can return to whenever you are slipping back into old habits. Remember, setbacks are a normal part of any change process, what matters is not that you never fall, but that you always get back up.

In the words of Carl Jung, "Who looks outside, dreams; who looks inside, awakens." This journey has been about turning our gaze inward, becoming more mindful of our actions, and reclaiming control over our digital lives.

As we step into a future that is increasingly dominated by screens, let us commit to being masters of our technology, not its slaves. Let us promise to value real-world experiences over digital ones and to prioritize human connection over online interaction.

Thank you for embarking on this journey. Here is to a future of digital freedom, mindful living, and enriched relationships.